FIRE TRUCKS

Illustrated with photographs

TRUCKS

Hope Irvin Marston

Dodd, Mead & Company

New York

The photographs in this book are used by permission and through the courtesy of: Boston Whaler, Inc., 41; Burch Mfg. Company, Inc., 23; W. S. Darley & Company, 10, 11, 12, 13, 21; Emergency One, Inc., 14, 15, 28, 29, 36, endpapers; Fire Department of New York, 6, 26, 32, 33, 35, 38, 39; Grumman Emergency Products, 22; Ladder Towers, Inc., 9; Los Angeles Fire Department, 40, 46, 47, 48, 55; Mack Trucks, Inc., 8, 18, 20, 24-25; Miami Fire Department, 42, 43; Oshkosh Truck Corp., 37; Philadelphia Fire Department, 27, 34, 50, 52; Pierce Manufacturing Inc., 19; Seagrave Fire Apparatus, Inc., 30, 31; Super Vac Trucks, 49; USDA/Forest Service, 44; *Western Fire Journal*, 16, 45; Wide World Photos, 17, 51, 53.

2 3 4 5 6 7 8 9 10

Library of Congress Cataloging in Publication Data

Marston, Hope Irvin.
Fire trucks.
Includes index.
Summary: Explains kinds of fire trucks, the equipment they carry, and the functions they can perform in fighting fires.
1. Fire-engines — Juvenile literature. [1. Fire engines]
I. Title.
TH9372.M37   1984   628.9′25   84-8068
ISBN 0-396-08451-6

86-553√

This book is dedicated with respect and affection to the fire fighters in my family... both past and present... Barney, Less, Dave, Bob, Skip, Rollie, Pete, and Billy

The author wishes to thank Bill Gardner, a fire fighter with the Watertown (New York) Fire Department, for his generous assistance in the preparation of this book.

Red lights flash...sirens wail...bells clang. A charging fire truck roars down the street. It screeches to a halt in front of a blazing building.

Fire fighters leap to the ground. They yank hoses from the racks. Each one hurries to do his job. The blaze is brought under control in minutes.

Fires are not always put out quickly. Many rage on for hours. They destroy lives and property. The cry, "Fire!" is frightening.

How lucky we are to have fire fighters. Every day they risk their lives to protect us from fires.

Fire fighters use different kinds of trucks. Pumpers are hose trucks with pumps. They arrive at a fire first. Their job is to get water on the fire quickly. Pumpers come in different sizes. They can pump hundreds of gallons of water per minute.

Ladder trucks carry ladders and equipment to rescue people from buildings. They do not carry water. Aerials are trucks with very long ladders that can be raised 100 feet or more. They are needed to fight fires in tall buildings.

Pumpers are "attack" vehicles. They carry their own pump, hose, a supply of water, and tools. Fire fighters on a pumper work fast to get water on the burning building right away. They hose it down and cool it off. Then firemen can work inside.

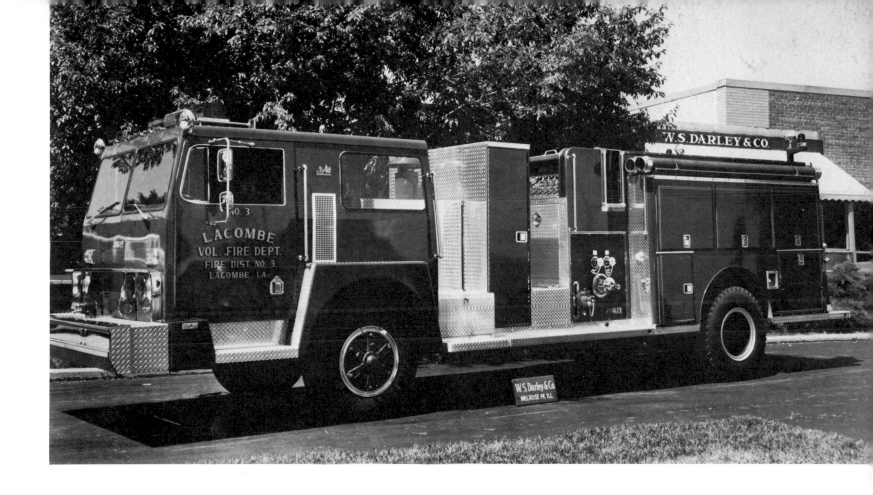

Pumpers can pump hundreds of gallons of water per minute. The force of the water is controlled by dials on the truck. Pumpers also carry ladders, hand extinguishers, and foam. They carry axes and pike poles for breaking through windows or walls.

While a pumper is emptying its own water tanks, hose is connected to the local water supply. If there is a hydrant nearby, soft suction hose that looks like gray canvas is used. When water is to be brought from a river or pond, hard suction hose is used. Those black "pipes" carried on a pumper are hard suction hose.

A hose weighs a lot when water is being pumped through it. It becomes as stiff and heavy as a steel pipe. That makes it hard to handle. It takes more than one fire fighter to hold it in place.

If the firemen let go of the hose when the nozzle is open, it could whip around. The force of the water shooting out could kill someone.

Ladder trucks carry different kinds of ladders. Wall ladders are held in place by hand. Extension ladders come in two or three sections. They are pulled out by pulleys and ropes. Roof ladders have hooks that fit over the peak of a roof.

The longest ladder is the aerial ladder. It rises right up from the fire truck. It can reach as high as ten stories or more.

The aerial ladder stays on the truck. It is operated by a motor. A fireman pushes buttons to raise it. The aerial ladder is on a turntable and can be moved around in any direction.

Aerial ladders help firemen reach high places. They are used to get people out of burning buildings.

The long ladders get men and equipment to the roof or to the upper stories of a building. Sometimes firemen have to cut holes in the roof to "ventilate" a fire—to let the smoke and heat out.

Fires can be fought from the air. Firemen use an aerial ladder with a platform at the top.

Minipumpers are small pumper trucks. Because of their small size, they can get through busy city streets fast. They pump 300 to 750 gallons of water per minute. By the time they empty their tanks, firemen arrive with the big pumper and lay hose line to the source of water.

Pumpers are custom-built to meet the needs of the fire department purchasing them.

When deciding what kind of pumper to buy, lots of things are considered—the weather, the kind of streets and roads to be traveled, the size of the community, the height of the buildings.

One company's trucks need extra protection from bad weather conditions. Another needs four-wheel drive because country roads are muddy in the spring. A third company requires extra hose and a deeper hose bed to carry it. It serves farms located far from a source of water.

Middle-sized pumpers are called midipumpers. Their pumps can deliver from 750 to 1,000 gallons of water per minute. Booster tanks can be added to carry more water.

Some models have an extra-short distance between the front and rear axles. That makes it easier to make turns in narrow streets.

Sometimes there is no water available where a fire breaks out. Then a tanker must carry water to the blaze. Firemen set up a portable water tank. The tanker dumps its load into it and rushes back for another. After the pumper at the fire empties its tank, it pumps water from the portable water tank.

Tankers carry 1,000 to 15,000 gallons of water. Some have pumps and hose, just as pumpers do.

The most powerful pumper ever built was the Super Pumper. It was one of five pieces of equipment that made up New York City's Super Pumper System (SPS). The SPS was used from 1965 until 1982.

The Super Pumper, nearly 44 feet long, was like a tractor-trailer. There was also a Super Tender and three Satellite Hose Tenders. The five units working together could pump 8,000 gallons of water per minute. That's 37 tons of water!

The Super Pumper and Super Tender did not have to be right at the scene of a fire. They could be at a water source as far as ten blocks away, hooked up to the Satellite Hose Wagons. Water cannons on the Tender and Satellites shot water 500 feet into the air.

Fire-fighting equipment like the Super Pumper was needed as taller and taller buildings were built. Taller ladders were also needed. Aerial ladders are 100 feet long or more. Some have platforms or buckets added to them.

The use of platforms on aerial ladders makes fighting a fire easier. Firemen can get a better view. They have a solid base on which to stand. They can rescue trapped victims more easily.

A Snorkel (opposite) has a basket platform, called a bucket, that can move up and down. A Snorkel is like a gooseneck crane. It works like a giant elbow.

Hose line runs up to the Snorkel's bucket. Fire fighters in the bucket can reach out and over roof-tops. They can be moved over, under, around, or between overhead wires or other obstructions.

Longer ladders meant that fire trucks had to be longer. They had difficulty turning corners. So a tiller cab at the rear of the truck was added. A tillerman steers the rear wheels. The driver steers the front wheels. The tillerman communicates with the driver by a bell, buzzer, or an intercom system.

The tillerman is like the end man in a crack-the-whip game as he guides the rear of the truck around corners.

Tiller cabs have windshields and seat belts to protect the tillerman. Some tiller cabs are open. But closed cabs are much safer.

Aerial ladders deliver water in different ways. Some have mounted hoses, and others have pipes, called standpipes, running up under the ladder. Still others have pipes that fold into each other like a telescope. There is a remote-controlled nozzle at the top. It is safer for firemen if they do not have to hold hoses in place.

Firemen at the top of an aerial ladder are in a dangerous place. Both smoke and freezing rain can be harmful to them.

Snorkels and aerials often work side by side.

Saving a life is the fireman's
first concern.

Airports use a special kind of fire truck—a crash
rescue vehicle, or CRV. These trucks are a special
kind of pumper. They carry foam and chemicals to
put out airplane fires. Usually they are painted
bright yellow. That makes them easy to see from
the air.

Airplane fires are often caused by fuel that has ignited. CRVs have roof-mounted turret guns. Chemicals are shot onto the fire as the CRV approaches. Haste is important. If the airplane's fuselage breaks in a crash, survival time is only seconds.

Fires on board ships or on docks or islands are best fought with fireboats. Large port cities have fireboats. They are like large tugboats with fire-fighting equipment. This includes scuba diving gear.

The world's largest fireboat is the *Fire Fighter*. It was built for the New York City Fire Department in 1938. It is still in use. It is docked at Staten Island.

The Los Angeles City Fire Department has five fireboats working in its harbor area. They handle any waterfront fires. Sometimes fireboats work with engine companies on land.

Not all fireboats are gigantic in size. The Boston Whaler is a small boat that can get around docks and harbors easily. When used as a fireboat, it is equipped with fire, rescue, and diving equipment.

The city of Miami, Florida, protects her seaport and her shorelines with a fireboat that can operate on land or on water. It has twelve hose connections. It can pump 3,000 gallons per minute. It pumps water to land-based pumpers or directly onto fires.

The *Fire-Rescue* can also be used as a rescue platform during hurricanes.

Not all fires are fought on land or on water. Some are attacked from the air. The U.S. Forest Service uses small planes to spot fires and decide where smoke jumpers should land.

Helicopters are valuable for fighting brush and forest fires. They make water drops, or drop chemicals. They lay hose lines up a steep hillside. They lower men and equipment into fire areas. And they rescue people from stricken areas.

In forest and woodland areas, bulldozers are part
of the fire-fighting equipment. They are used to
remove brush and grass and make a firebreak, so that
the fire will die out when it reaches bare ground.
They are also used for building and maintaining
fire roads.

A heavy utility tractor is used by firemen to lift heavy objects. It pulls down the remains of burned-out buildings. And it carries special equipment for rescues at cave-ins, auto accidents, and airplane crashes.

Fire fighters who must enter smoke-filled buildings wear air tanks that are connected to face masks. So do scuba divers. When these air tanks are empty, they must be refilled with compressed air.

When large quantities of air are needed, a fire department may have an air utility wagon. This provides a refilling station for air tanks at a fire. Some companies use a smaller, portable unit that will fit into the bed of a pickup truck.

Lighting trucks aid in fighting fires in the dark. They help firemen to see exactly what is happening. This one can give enough light to light up an entire football field.

It takes more than good equipment to fight fires. Brave men and women are needed to operate the equipment.

At times it takes teamwork to rescue trapped victims and carry them to safety.

Fire fighting is necessary, no matter what the weather is like.

It is not just a job for men. More and more women are joining the ranks of professional fire fighters.

These men of the Los Angeles Fire Department represent a complete line of defense. Pictured in their working uniforms, they serve on land, sea, and in the air.

With trained, able fire fighters like these, and modern up-to-date fire trucks and equipment, we can sleep well at night. They will protect us from fire.

Index